not
the L
dema
PLE

26

ENERGY

PETER D. RILEY

First published in Great Britain by Heinemann Library
Halley Court, Jordan Hill, Oxford OX2 8EJ
a division of Reed Educational and Professional Publishing Ltd.
Heinemann is a registered trademark of Reed Educational &
Professional Publishing Limited.

OXFORD FLORENCE PRAGUE MADRID ATHENS
MELBOURNE AUCKLAND KUALA LUMPUR SINGAPORE TOKYO
IBADAN NAIROBI KAMPALA JOHANNESBURG GABORONE
PORTSMOUTH NH CHICAGO MEXICO CITY SAO PAULO

Designed by Visual Image
Printed in Hong Kong

02 01 00 99 98
10 9 8 7 6 5 4 3 2 1

ISBN 0 431 08435 1

British Library Cataloguing in Publication Data

Riley, Peter, 1947-
    Energy – (Cycles in science)
    1. Force and energy - Juvenile literature 2. Energy
    development - Juvenile literature 3. Renewable energy
    sources - Juvenile literature
    I. Title
    333.7'9

Acknowledgements

The Publishers would like to thank the following for
permission to reproduce photographs:
Planet Earth Pictures: P David p10; Science Photo
Library: Mary Evans Picture Library: p5 M Bond p19,
26, J Charmet p11, Will & Dent McIntyre p24, P
Plailly p29, R Ressmeyer/Starlight p25, Science
Museum/Science & Society Picture Library pp 12,
18, T Takahara p28, US Department of Energy p22;
Tony Stone Images: F Herholdt p13, P O'Hara p17, P
Seaward p27, R Wetmore II p6, K Wood p14; Zefa:
pp4, 7, 8, Science Photo Library p23, V Leeden p21.

Cover photograph reproduced with permission of
Novosti Press Agency/Science Photo Library.

Our thanks to Jim Drake for his comments in the
preparation of this book.

Every effort has been made to contact copyright
holders of any material reproduced in this book. Any
omissions will be rectified in subsequent printings if
notice is given to the Publisher.

Any words appearing in the text in bold, **like
this**, are explained in the Glossary.

CONTENTS

ENERGY

*Find your **pulse** and feel the movement of the artery. It is caused by your beating heart which is using energy deep inside your chest to keep you alive. Listen to the sounds around you. They are made by energy moving through the air and shaking your **eardrums** backwards and forwards very quickly. Electricity and fuels like gas provide us with energy too. Saving energy in fuels now may help generations of people in the future.*

WHAT DOES ENERGY MEAN?

This surfer is using the moving energy in a huge wave to race up to the beach.

Energy seems like a modern word but it has been used by scientists for nearly 200 years. Thomas Young, an English scientist living in the early 19th century, used the word energy to mean 'the ability to do work'.

WHERE DOES ENERGY COME FROM?

Scientists believe that everything in the universe today – the **materials** and the energy that make things work – came from a huge explosion they call the Big Bang. This happened 15,000 million years ago. Since then, the universe has been expanding.

WHAT KINDS OF ENERGY ARE THERE?

There are many different forms of energy, and any one form can change into another. Almost everything has energy stored in it. This is called **potential energy**. When stored energy is released it may change into heat energy, light energy, sound energy or moving energy, called **kinetic energy**.

The potential energy which this tortoise had (when it was in the eagle's beak) is changing into kinetic energy as it falls through the sky. It might also change into some sound energy when it hits the man's head!

WHERE DOES ENERGY GO?

All energy eventually changes into heat energy which spreads throughout the universe. Some scientists believe that this will continue for ever until the universe eventually cools down (as the universe is expanding it is cooling down). Other scientists think that the universe may end with a Big Crunch, when everything in it is squashed back together and the energy will recycle again in another Big Bang.

CONSERVATION OF ENERGY

Energy cannot be destroyed but it can be changed into other forms. This is what a scientist means by energy conservation, but there is another meaning: saving energy, such as switching lights off in the home when nobody is using them.

MOVING ENERGY

Energy moves objects we can see as well as the tiny particles that make up the objects which we cannot see. Energy also travels as waves of electricity and magnetism.

MOVEMENT EVERYWHERE

Every movement we can see from a plodding tortoise to a jet plane screeching across the sky is due to **kinetic energy**. It will move your eyes as you read this page and move your chest to keep you breathing.

THE PARTICLES OF MATTER

These hot air balloons inflate because the heaters below them give so much energy to the air **particles** inside that they move quickly and push outwards on the balloons' sides.

Solids, liquids and gases are made of tiny particles we cannot see. In a solid the particles are stuck together but make tiny movements called vibrations. In liquids the particles can move around each other and in a gas they move away from each other like the hot grains of corn in a popcorn maker. Cold chocolate is a solid but when it receives heat from your hand it turns into a liquid. The heat energy it receives turns to kinetic energy and makes the particles move round each other.

ELECTRIFYING

In metals there are **electrons** which can move and make a **current** of electricity. If the metal is connected to a battery, electrical energy pushes the electrons one way around the circuit. In a power station the electrical energy is produced by a **generator**. It makes the electrons move backwards and forwards along the cables to your home through the wires in your mains and into lamps, heaters and electrical equipment such CD players, television sets and computers.

THERE IS NOTHING FASTER

In the 19th century it was discovered that when electricity flowed through a wire it made the wire behave like a magnet and turn a compass needle. Another experiment showed that moving a magnet past a wire made a current of electricity in the wire. Later experiments showed that electricity and magnetism form waves called **electromagnetic waves** which can travel through air and space. The heat and light energy we feel and see are electromagnetic waves moving at nearly 300,000 kilometres every second – the fastest speed anything can move. There are many kinds of electromagnetic waves.

The electrical energy released from this storm cloud makes a zig-zag path through the air. Some of it changes to light, heat and the sound of thunder as it makes its journey to the ground.

ENERGY STORES

*Energy is stored in many different ways. Stored energy is called **potential energy** because it has the potential or capability to change into another form of energy.*

GOING DOWN

If you put a toy car at the top of a ramp and let it go it rushes to the bottom and across the floor. There was no motor in the car and you did not give it a push, but the car suddenly released enough energy to send it whizzing along. The energy was stored energy due to **gravity**. Everything that can fall to the ground has this gravitational energy stored inside it.

IT'S A WIND UP

Springs and elastic bands can store energy. If you wind up a clockwork toy a metal spring inside becomes **coiled** and stores the energy. When the toy is released the metal spring uncoils and the potential energy changes to movement and sound energy.

When this **propeller** is released the energy stored in the twisted elastic band will make the band uncurl, spin the propeller and push the model forward through the air.

CHEMICAL STORES

There are thousands of different chemicals and they all have energy stored in them. Two high energy food chemicals are **carbohydrates** found in potatoes and rice, and **fats** found in butter and oils. Your body uses their energy to help keep you alive. The batteries in a torch are stores of chemical energy.

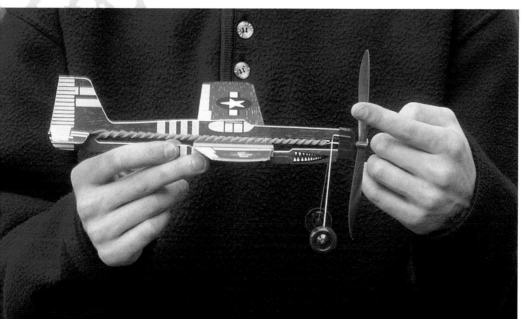

When you switch on
the torch the energy is released
to drive **electrons** through the wires. When they reach
the thin wire in the bulb, called the filament, some of
their energy changes to light.

NUCLEAR ENERGY

Solids, liquids and gases are made of **atoms**. Each atom
is like a ball. At the centre of an atom is the **nucleus**. It
contains energy. A number of atoms have such a large
nucleus that they cannot hold onto all their energy and
part of it is released as **electromagnetic waves**. Some of
this is in the form of heat energy which can be used in
power stations to make electricity. Other electromagnetic
waves, called gamma **radiation**, which are also released,
are harmful if they are allowed to pass through living
things. Power stations using nuclear energy are designed
to prevent harmful radiation escaping into the
environment.

A few
moments before
this photograph was
taken these firework
rockets were in their
launch tubes. When their
fuses were lit the energy
stored in the chemicals in
the rockets was released
and the rockets shot into
the air to make this
display.

ENERGY AND CHANGE

In the past people thought that a fluid made things hot. Today we know that heat is not a fluid, but is the result of one of many energy changes. Energy changes can be linked together to make an energy chain.

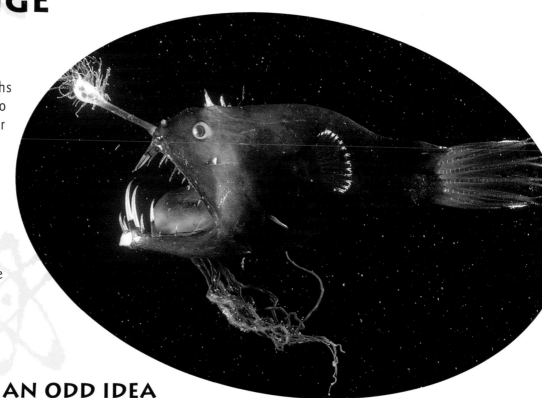

In the ocean depths there is no light so the fish make their own. The light energy in the lure of this angler fish is being used to attract other fish to eat. They will provide more energy to keep the lure lit.

AN ODD IDEA

In the 18th century scientists believed that heat was a fluid that could be found in everything. They thought that when things became hot it was due to the fluid escaping through tiny holes in the surface of the substance. This fluid was given the name caloric, and when a hot drink in a cup warmed your hand it was thought to be due to the caloric passing in through your skin from the side of the cup.

BOILING WATER WITHOUT A FIRE

In 1789 an American scientist called Count Rumford was making a cannon by drilling out the centre of a metal barrel. As the cannon was being made by the turning action of the drill the metal became so hot that it had to be cooled with water.

After two and half hours the water around the cannon had become so hot that it boiled. He worked out that the cannon could not have contained all the caloric needed to make the water boil, as it would have melted straight away. He decided that the heat was a form of energy that had come from the moving energy of the drill.

A CHAIN OF ENERGY

If you shut this book quickly you will hear a sound. Some of the energy stored in your food changed to movement energy in your muscles, then some of this energy changed to the movement energy of the book covers, and finally some of this energy changed to sound energy. This path of energy is called an energy chain.

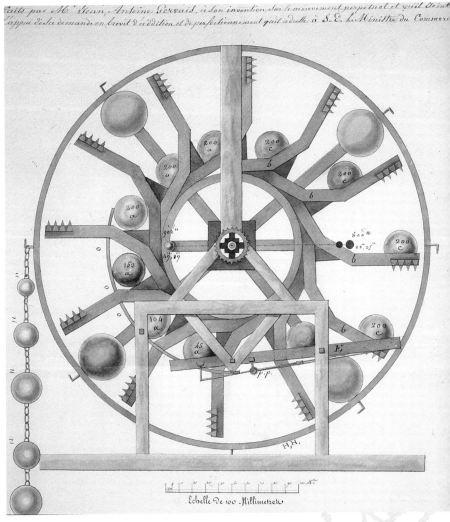

This is a perpetual motion machine. It is designed to run for ever but it never will. Some of the friction between the rolling ball and the groove releases heat energy which is lost into the air.

11

HEAT ENERGY

Heat energy is released at every energy change. It can move in three different ways. It is measured using thermometers that are marked with a scale of temperatures.

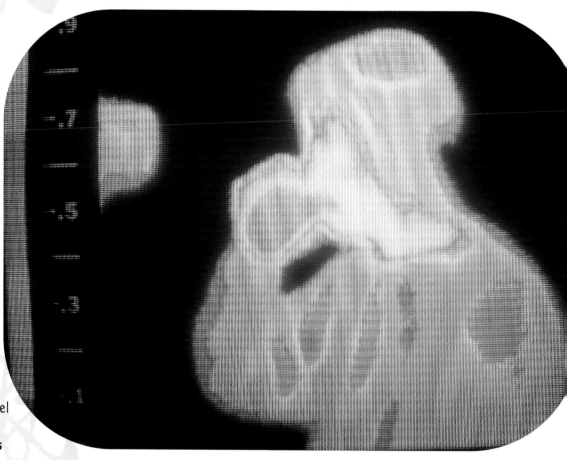

Heat can travel by **radiation**. **Infrared rays** pass out from warm bodies into the cooler air. These rays can make a photograph on special film which shows the hottest (red and yellow) and coldest (blue) parts of the body.

HOTTING UP

Should you be somewhere else? Are you late? If you are you will shut this book and run. After a few minutes running you will feel yourself getting hotter. The chain of energy in your action starts with the stored energy in your food... which is released to move your muscles... that move your leg bones... that move your body to where it should be. When energy changes from one form to another most of it is converted into heat. That is why you get hotter when you run.

HEAT ON THE MOVE

Heat only moves from a hotter place to a cooler place. It moves in three different ways. Heat can move as infrared rays that travel at **light speed** through air and space like the Sun's rays that warm us. It can move by **conduction** in a solid like a metal spoon left in hot soup that gradually gets too hot to hold. In conduction think of the heat as being in parcels and the metal passes parcels of heat along as in a party game. Heat can also move by **convection** where the substance with heat carries it away rather like someone running away with the parcel and spoiling the passing game.

THERMOMETERS AND TEMPERATURE

The Italian scientist Galileo invented the thermometer to measure temperature in 1593. It was filled with air. Later thermometers had a liquid inside them like the ones we use today. When the liquid is cold it **contracts** and when it is warm it expands. In the 18th century the temperature scale we use today was developed.

The metal in this wok conducts heat quickly from the flame to the food. The wooden handle does not conduct heat well so the cook's hand does not get burned.

ENERGY USES

We use so much energy we rarely think about it. In the first hour we are awake we use many different forms of energy and we continue to do so for the rest of the day. Energy is used to make all the things we need.

On this production line cars are assembled from metal, plastic and glass. Energy is used by the power tools to join the parts together and move the cars along to the next set of assemblers.

AN ENERGETIC DAY

Did an alarm clock wake you up this morning? Perhaps it played a tune or made the radio come on. Electrical energy changes to sound which tells you it is time to get out of bed. Hot water splashes out of the tap in the bathroom. Energy in gas, oil or electricity has been changed to heat so the water can warm and clean your skin this morning. Into the kitchen and on with the kettle and maybe the toaster too. Electrical energy supplies heat energy straight away to make your breakfast. An egg boiling in a pan of water on a gas stove takes in heat to change its runny insides to a white and yellow solid. On the way to school cars roar by, converting the stored chemical energy in petrol into movement energy. How do people use energy throughout their day? Find some answers in the upside-down box below.

Answers: computer, television, radio, video, CD / cassette player, making things in factories, transporting things in trucks, ships and aircraft.

MADE BY ENERGY

Look at your clothes. They would not exist without energy, which has been used to make them. The fibres may have grown on the back of a sheep. The sheep used energy from its food, which came from sunlight trapped by grass. Energy was also used by machinery to put the fibres together into the cloth that covers your skin. It was used to make the dyes which colour your clothes too.

Energy was used to make your favourite mug, the glass in your bedroom window and all the goods in your home, from a safety pin to a television set. It has been used to make the bricks and blocks in the walls of your home and the tiles on the roof. Vast amounts of energy are released to put together all the materials you use in your life.

Heat energy has been used to release iron from its rocky **ore** and change it into this **molten** steel, ready for making it into a car.

THE SUN GOD

Our ancestors used to worship the Sun as a god. Scientists today are attempting to mimic the Sun and make energy in the same way as the Sun, here on Earth.

THE POWERFUL SUN

Some of the energy in sunlight shining into this wood is trapped in the leaves. It is used by the plants to make food from air, water and minerals in the soil. Some of the plants' food is eaten by the animals living in the wood.

From the earliest times people have marvelled at the power of the Sun. When it was low in the sky the weather was cold, food plants died, game animals perished, moved away or were difficult to find and people starved. When the Sun was high in the sky the weather was warm, food plants grew, game animals bred and there was plenty of food for everyone.

JUST FAR ENOUGH

The Earth is just at the best possible distance from the Sun for the Sun's heat to provide energy for life. If the Earth was closer, the Sun would provide too much heat for living things to survive. If the Earth was further away it would be too cold for life.

WHERE THE SUN'S ENERGY COMES FROM

The amount of **material** or matter in a substance is called its mass. Mass is a very concentrated form of energy. In the Sun four million tonnes of matter are destroyed every second as the force of the Sun's **gravity** squashes up hydrogen gas and makes helium gas. The energy released keeps the temperature at the Sun's **core** at 15,000,000 °C.

COPYING THE SUN

The inside of the sun is made from a substance called plasma. Energy is released from it. On earth, scientists have invented a machine called a tokamak in which plasma can be made. It is hoped that this will be able to produce useful energy.

Note: Investigations of the Sun are made by using special equipment. Never look directly at the Sun as it can harm your eyes.

Most of the surface of the Sun is 5800 °C but in some places it is cooler. These places make dark patches on the Sun's surface and are called sunspots although they may cover an area of 2000–3000 km^2. Sun spots may last up to a few months.

CAN WE USE IT AGAIN?

Some sources of energy were formed so long ago that they can only be used once and cannot be reused. Other sources of energy can be used again and again.

OLD ENERGY

About 275 million years ago large forests of evergreen trees and giant ferns grew on swampy land. When the forest plants died and fell into the swamps their remains were **preserved** and became covered with rock. Eventually, they changed to coal. The energy taken in from the Sun by these ancient plants was stored in the coal. Today coal provides the energy for many power stations. The light in your home may be releasing energy first trapped from sunlight before the time of the dinosaurs.

Oil and gas are made from the bodies of tiny sea creatures that died 200 million years ago. These fuels, along with coal, made from the preserved parts of living things, are called fossil fuels. When we have used them up they will not be replaced. They are non-renewable energy sources.

The light energy is released when the oil burns at the top of the wick, while the level in the reservoir falls. Oil burning power stations are using up the world's oil resources in a similar way.

RENEWABLE RESOURCES

In many parts of the world wood is cut and used for fuel. Some people collect cattle dung and dry it to make fires. When these fuels run out there is always more to collect.

This experimental machine uses energy from waves to make electricity. If it works successfully a machine fourteen times larger will be built to supply power to local villages.

More energy reaches the Earth from the Sun in a week than is stored in all the fossil fuels. Light energy can be trapped by **solar** panels on **satellites** and pocket calculators and changed into electricity. Most of the Sun's energy moves the air as winds and **evaporates** water from the oceans. The moving energy of the wind can be converted to turning energy in a windmill to make a pump work or produce electricity. When water in the air **condenses** and falls as rain it can be trapped behind a river dam and released through a chamber to spin a **turbine** like a water wheel. This is connected to a **generator** that converts this movement into electrical energy. Scientists are working on ways to use more energy from sunlight, wind and moving water.

CHANGING THE WORLD

People used the energy in animals, the wind and moving water long ago. When the energy in steam was discovered it changed how things were made and how people travelled.

USING ENERGY LONG AGO

The first people used only the energy in their own bodies to hunt, collect plants for food and make clothes from skins. Later the ox and horse were domesticated and their energy was used to pull ploughs and carts. The heat energy in fire was used to clear patches of forests in order to grow crops. In some parts of Australia people used fire to hunt animals. In Roman times the energy of moving water was used to turn wheels in water mills. This energy was used to grind corn, make paper or clean cloth. By 644, in places where there was little water, the moving energy of the wind was used to turn the sails of windmills and grind corn.

The steam is made in the long boiler above the wheels. It passes to the **pistons** in the tubes with white ends and pushes them backwards and forwards to turn the wheels.

GETTING STEAMED UP

In 18th-century England people were running out of wood to burn. It had been discovered that coal could be used as an alternative but it was dangerous to collect because the mines often became flooded. A steam engine was invented to pump water out of the mines. James Watt, a Scottish engineer, improved the design of the steam engine and made it more powerful.

The engine had a moving part called a piston. It moved backwards and forwards. Watt made attachments for the piston so it could turn a wheel. This made it more useful for working all sorts of machines.

THE INDUSTRIAL REVOLUTION

Before the invention of the steam engine almost everything was made by hand or by people working machines, like a weaving loom, themselves. Once the energy in moving steam could be used to turn a wheel, **belts** were connected to weaving looms and spinning machines and driven by steam engines. The engines were so powerful that they could work hundreds of machines at once. Mills and factories were set up to house the steam engines and the machines they worked. Workers were needed to tend the machines and deal with the raw **materials** and products that were made, and so towns developed around the mills and factories. In less than a hundred years industrial towns had been set up all round the world to make a wide range of products more quickly and cheaply than ever before.

A belt being used to transfer the moving energy from a steam engine (not shown) to this machine (which extracts juice from sugar cane).

USING ELECTRICITY

The electrical energy generated in a power station is changed into many other forms of energy when it reaches our homes.

THE ELECTRICAL ENERGY CHAIN

This is an electricity **generator** in a power station. You can judge its size by using the man on the left as a reference. Inside the yellow casing, steam spins the turbine and magnet.

Electricity is generated in a power station. The source of stored energy at the beginning of the chain may be coal, oil or nuclear energy. The most useful energy released from fuel is heat. It is used to raise the temperature of water until the water turns into steam. The moving energy of steam is directed through gaps in **turbine** blades which makes them spin. A magnet is attached to the turbine and they spin together. Around the magnet are huge **coils** of wire. As the magnet spins its magnetic field passes through the wires and pushes and pulls on the **electrons**. This makes them move backwards and forwards along the wires, through the cables from the power station to your home. This is electricity, which comes through the wires in your mains every time you switch on an electrical appliance.

TUNING IN

When you switch on the radio, the energy in radio waves from the local **transmitter** makes a **current** of electricity which moves the speaker to make a sound.

In a kitchen electrical energy is used in the lights, fridge, air extractor, dishwasher and food processor.

IN THE PICTURE

If you press the television remote control, the stored chemical energy in the batteries changes to electrical energy. This changes to waves of **infrared radiation** which cross the room and activate the standby switch. Electrical energy from the mains passes through the electronic components of the television set. Behind the screen is a **cathode ray tube**. In it electrical energy is changed to heat energy which changes to movement energy that shoots beams of electrons onto the back of the screen. More electrical energy moving through a set of coils makes these generate a magnetic field which pulls the electron beam in a zig-zag down the back of the screen. The energy in the moving electrons is changed to light energy as they hit chemicals, called **phosphors**, on the glass screen and we see the pictures.

MAKING A MESS

We use fuels to release energy, but there is a risk that we may damage the environment in a variety of ways.

IN HOT WATER

If hot water from a power station is released into a river the heat energy warms the water and speeds up the life processes of water plants. They grow and breed quickly and turn the water green. When the plants die the bacteria feeding on their bodies take all the oxygen out of the water. This starves fish and other water life, like snails, of oxygen and they die.

These trees were killed by air pollution and acid rain that was made when energy was released from fossil fuels to make steel and oil products in factories.

ACID RAIN

As the heat energy is released from a burning fossil fuel, such as coal, sulphur is also released. This forms sulphur dioxide gas as it passes out of the power station chimney in the smoke into the air. When sulphur dioxide mixes with the water vapour in clouds it forms **sulphuric acid**, which eventually falls to the ground in raindrops. The acid reaches the soil where it can cause trees to die and if it reaches streams and lakes it can kill water life too.

MORE THAN HOT AIR

There are **infrared rays** in the light that reaches us from the Sun. These rays pass through the atmosphere of the Earth and heat the ground. The warm ground sends infrared rays back into the atmosphere and space. Power stations and the exhaust fumes of cars and trucks around the world add nearly 1000 tonnes of carbon dioxide to the atmosphere every second. Like the glass in a greenhouse, this gas traps the infrared rays and stops them leaving the Earth. This causes the atmosphere to become warmer. This warming of the planet is called **global warming** and is said to be due to the **greenhouse effect**.

IS IT WORTH THE RISK?

Some power stations use nuclear fuels which are **radioactive**. They release large amounts of heat but also release harmful fast moving **rays**. The nuclear fuel is kept safely at the centre of the power station but accidents have happened and **radiation** has escaped with fatal results, also causing damage to animals and plants in the environment.

When radioactive materials are being handled, special equipment and clothes must be used to protect the worker from the harmful high-energy radiation.

CHANGING WAYS

Every way of generating energy is now being made less harmful to the environment. Hot water is being re-cycled. Sulphur dioxide is being removed from power station smoke. Alternative ways of generating energy without releasing carbon dioxide, such as using wind power, are being developed.

HOME AND AWAY

Energy can be saved in the home, in the way we get rid of our rubbish and in the way we travel from place to place. Saving energy saves more non-renewable fuel for the future and reduces pollution and the risk of global warming.

Warm air rises to the ceiling and moves by **conduction** into the loft and from there to the roof. Covering the floor of the loft with insulating blankets keeps the home warmer so the heating system can be turned down or off.

THE GREAT ESCAPE

Most of the energy released in the home is used to keep it warm and to heat water. More than a third of the heat leaving a home escapes through the walls. A quarter of the escaping heat goes through the roof and the rest escapes through window glass, floors and draughty doors.

Air is a poor conductor of heat. This means that if air is trapped the heat passing through it will slow down. In a **double-glazed** window the air is trapped between two sheets of glass. Thick blankets trap air between their fibres and are used to stop heat escaping from a roof or from a hot water tank.

The simplest way to prevent heat escaping from your home is to make sure all doors and windows have draught excluders which fill any gaps where heat may pass through to the outside.

HIDDEN HEAT LOSS

Most of the items you buy and bring home are enclosed in packaging. A great deal of energy is used to make the bottles, cans and cardboard that are thrown away every week. They can be taken to a recycling centre and the **materials** processed into new items. Less energy is used in recycling than in processing raw materials.

CHANGING HABITS

More energy could be saved and the **greenhouse effect** could be reduced if people used their cars less. It would be healthier to walk or cycle for short distances, and for long distances it would be better to use public transport. A train or a bus moves many more people than a car.

Traffic in towns and cities spends more time in stationary queues than moving along the roads. The idling engines waste energy and pollute the air with their exhaust fumes. Well-planned public transport moves people faster and uses less energy.

THE FUTURE

Every day more and more energy is used by the growing human population. Each day the amount of fossil fuels gets lower and lower. Oil and gas will be used up first. Coal may only last a few hundred years. Other sources of energy will have to be found to replace the fossil fuels, and these will have to be used wisely so that as little as possible is wasted. This is being tackled in a variety of ways.

Energy is lost when a car pushes through the air. By studying the **air flow** over a car body new designs can be made which reduce energy losses to a minimum.

HEAT TRAPPING

Solar panels are like flat greenhouses that trap the Sun's heat. Water flows through pipes in the solar panel and carries the trapped heat away. Solar panels can be placed on the roof of a house and connected to the heating system.

WIND AND WATER

The moving energy in wind and water can be used to turn **generators** and make electrical energy. **Turbines** on tall supports are being built on windy hillsides. Their blades are designed to turn in gentle breezes so that they can produce some electricity on most days of the year. The up and down movement of the waves at sea could provide energy for making large amounts of electricity in the future.

NUCLEAR ENERGY

Nuclear power stations have the advantage that they do not pollute the environment with harmful gases, but the waste from nuclear fuel is so **radioactive** that it has to be stored for thousands of years before it will become harmless. If a better way of treating the wastes could be developed, more nuclear power stations could be used.

PLANT POWER

In developing countries, where wood is the main source of energy, huge numbers of trees will have to be planted and the cutting of fuel wood will have to be carefully controlled in order to keep a constant supply of energy for millions of homes.

Scientists are making new kinds of plants by genetic engineering. They can be grown to produce a substance to replace oil when the supplies of fossil fuels run out. In Brazil sugar cane is used to make the fuel to power cars. Perhaps in the future, scientists will find even more ways to use plants in the production of energy.

These solar cells are being tested to see how well they change light energy into electrical energy. Artificial rain can be made in this room to see how well the cells work in a shower.

GLOSSARY

air flow the movement of air over a surface

artery a tube which carries blood away from the heart to organs of the body

atom a very small particle of a substance

belt a band of material that forms a circle

carbohydrate a chemical which provides the body with energy

cathode ray tube a glass tube with a screen at one end and a device at the other that produces a beam of electrons, this zig-zags across the back of the screen, making a picture on the front

coil a wire wound into circles which are arranged next to each other

conduction the movement of heat or electricity through a substance

contract to become shorter or smaller

convection the movement of heat in a liquid or a gas

core the centre of an object

current the flowing movement of a gas, a liquid or electricity

double glazing a window made from two panes of glass with one in front of the other and a small air gap between them

eardrum a delicate membrane in the ear which vibrates when sound waves reach it

electromagnetic wave a beam of energy that can pass through air and space; some carry the signals that make radios and televisions work

electron a very tiny part of an atom which has a negative electric charge

evaporate to change from a liquid into a gas or vapour

fat a food substance which stores energy in the body and forms an insulating layer under the skin of animals keeping them warm

fuse the joining of substances that have melted together

generator a machine which makes or generates electricity

global warming the heating up of the planet and its atmosphere which brings changes to the weather and climate everywhere

gravity a force which pulls everything towards the surface of the Earth and acts as a pull of attraction between any two objects in the universe

greenhouse effect the raising of the temperature of the atmosphere by gases such as carbon dioxide which prevent heat from the Sun being reflected from

the Earth into space

infrared electromagnetic waves that carry heat energy; they are longer than light waves but shorter than radio waves

kinetic energy the energy in a moving substance or object

light speed the speed at which light travels; about 300,000 kilometres per second

material any substance which is made of matter; for example a solid, liquid or a gas

molten the condition of a solid that has been turned into a liquid by heat

nucleus the centre of an atom around which the electrons move

orbit the path taken by the planets around the Sun, or the moons around the planets or artificial satellites around the Earth

ore a rock from which a metal can be extracted usually by heating

particle a very small object ranging in size from a grain of sand to an object inside an atom such as an electron

phosphor a chemical which releases energy as light when electrons strike it

piston a disc which can be pushed backwards and forwards inside a cylinder and can transfer its moving energy by a series of rods to wheels to make them turn

plasma a hot gas in which the electrons have separated from their atoms

potential energy the energy in a substance because of its position relative to other things

preserve to keep the same

propeller an object with two or more blades attached to a central shaft, which turns to drive aeroplanes and ships

pulse a throbbing of an artery in time with the heart beat

radiation energy or particles released by a radioactive substance

radioactive the state of a substance that is breaking down on its own into another substance

ray a narrow beam of light or other form of energy

satellite an object which moves in an orbit around a planet such as a moon; also a machine for transmitting television programmes

solar relating to the Sun

sulphuric acid a corrosive liquid which is harmful to life

thermometer a device for measuring temperature

transmitter an electrical device which sends out radio electromagnetic waves

turbine a device with blades which are turned by air, steam or water; it is often connected to an electricity generator

INDEX